The Belgian Hare For Pleasure and Profit
A Textbook on the Belgian Hare

by U.G. Conover

with an introduction by Jackson Chambers

This work contains material that was originally published in 1915.

This publication is within the Public Domain.

This edition is reprinted for educational purposes
and in accordance with all applicable Federal Laws.

Introduction Copyright 2017 by Jackson Chambers

Self Reliance Books

Get more historic titles on animal and stock breeding, gardening and old fashioned skills by visiting us at:

http://selfreliancebooks.blogspot.com/

Introduction

I am pleased to present yet another title on Poultry.

The work is in the Public Domain and is re-printed here in accordance with Federal Laws.

As with all reprinted books of this age that are intended to perfectly reproduce the original edition, considerable pains and effort had to be undertaken to correct fading and sometimes outright damage to existing proofs of this title. At times, this task is quite monumental, requiring an almost total "rebuilding" of some pages from digital proofs of multiple copies. Despite this, imperfections still sometimes exist in the final proof and may detract from the visual appearance of the text.

I hope you enjoy reading this book as much as I enjoyed making it available to readers again.

Jackson Chambers

CONTENTS

	Page
Preface	3
Chapter I—Belgian Hares and what they are	4
Chapter II—Multiplication	5
Chapter III—Houses and Hutches	7
Chapter IV—Feeding	10
Chapter V—Mating and Breeding	11
Chapter VI—The Young, and How to Care for Them	14
Chapter VII—Raising Belgian Hares as a "Side Line"	16
Chapter VIII—Diseases and Remedies	18
Chapter IX—Young Folks with Belgian Hares	20
Chapter X—How to Cook the Belgian Hare	23
Chapter XI—Why Belgian Hares are more Profitable than Poultry	26
Chapter XII—A Talk with the Beginner in Belgian Hares	28
Chapter XIII—The Belgian Hare Business and its Outlook	29
Chapter XIV—The Record, and How to Keep it	32
Chapter XV—The Pleasant Ridge Rabbitry	32
Chapter XVI—Profit and Raising Belgian Hares	35
Chapter XVII—American Standard of Excellence	36

THE AUTHOR.

Preface.

THIS little book is presented to the public in response to a demand for information upon the subject of raising Belgian hares. It is intended especially for the beginner, and the aim of the editor has been to present only those things of tested worth. Exaggeration of the possibilities of the Belgian hare will not be found in this book. By a careful and thoughtful study of this work, any intelligent person may succeed in breeding Belgian hares.

Belgian hare raising is no longer an experiment; it is an industry that is just as legitimate and honorable as raising poultry or other live stock.

Notwithstanding the proverbial slander that the American people like to be humbugged, they are accredited as being the shrewdest and farthest-seeing people on the face of the earth, and when they undertake to accomplish anything they usually succeed.

The Belgian hare industry is now within their grasp, and is suitable for the millionaire and laborer alike as a side line, whether it be for pleasure or profit. The merchant and his clerk, doctors, lawyers, ministers, farmers, as well as their wives, sons and daughters—everybody, in fact, but the sluggard, the fop, and the "Weary Willie"—may find raising Belgians an interesting, as well as profitable, business.

Cordially yours,

THE AUTHOR.

384846

CHAPTER I.

Belgian Hares and What They Are.

THE Belgian hare of today is like the fancy poultry we see in most fanciers' yards—they have been bred up and conform to a certain standard, termed the Standard of Excellence.

Between the Belgian hare and the common rabbit we see running wild, there is just as much difference as day and night.

The wild rabbit seldom attains a weight of over three pounds, whereas the Belgian when mature will weigh from eight to twelve pounds.

Their meat compares favorably with frog legs—so many judges assert who have tested it out—it being more easily digested than many other meats.

The Belgian hare is said to have originated in Belgium, probably about the beginning of the nineteenth century, where it is now found small in size but perfect in form, color and markings. The modern Belgian hare is an animal of singular charm and great utility, combining the beauty and toothsomeness of the old domestic hare with the grace and fecundity of the wild rabbit, through a process of breeding that has been practiced for the past fifty years or more.

From the best data we can get, the Belgian was introduced into America about 1860, but its merits were little known, so it was by no means the perfect animal we see in the hutches of breeders and fanciers of today. It is only the last few years that the public is slightly awakening as to the possibilities of the Belgian.

That they are much more profitable than poultry the writer has proven beyond a doubt, but the ideal of the Belgian is that of its side line qualities. When Belgians are raised in connection with poultry it is an ideal combination, as the hen will lay the eggs and the Belgian will make the meat.

As the Belgian can be very satisfactorily bred four times in a year, producing from six to twelve to a litter, anyone with a pencil can soon figure up what he will have at the end of the year.

They do not require expensive feeds, and as once a day is sufficient for feeding, it is quite evident the care and feeding proposition is small. Alfalfa hay is an ideal ration with a handful of oats once a day and water. For anyone who can raise or purchase alfalfa hay the problem of feeding is solved.

They will eat clover hay as well as timothy, but timothy is a poor substitute for either of the other two just mentioned. Corn, carrots, cabbage, dandelions and many other things such as the sheep like are good for the Belgian, and he will thrive upon them.

The color of the Belgian is described as a "rufous red." It is rather hard to describe, being something of a fox color or deep golden tan. It is not distributed equally, but is richest on the shoulders and top of neck. The hair is tipped with black, called ticking, and the proper distribution of the ticking adds greatly to the beauty of the animal. The finest bred hares, or exhibition stock, must have four red feet and other qualifications to stand the least show in winning.

There is no place in America the Belgian will not thrive if cared for intelligently, and it does not require an expert to feed them, as the writer has many customers among boys and girls going to school who are quite successful in raising them and make considerable money as a side line. Have shipped hares as far north as Alaska, and the customer writes that the climate seems ideal for them; and on the other hand, the same story comes from as far south as Cuba. So no one contemplating starting need to hesitate on account of their location or climate.

Because of the present high cost of living the Belgian will certainly appeal to the laboring man, and there are people in every vocation of life who purchase Belgians and raise them to cut the cost of the meat bill. You not only get very cheap meat, but the very best the market affords. To some people who think the Belgian is not fit for eating in the summer, let me say in this connection that the writer, many years ago, when first starting in the business, sold large quantities to the very best hotels in Cincinnati in the month of July, and it can not be questioned but they knew they were just as eatable in July as December.

I have understood that in some parts of the country the fur of the Belgian is substituted for quite expensive furs, and the customer who purchases them pays a good, big price for them.

To anyone who has office employment, the getting out once or twice daily and attending to the wants of the hares, in the fresh air, getting their minds off the regular work, is good for the mind and body attending to their simple wants.

The Belgian hare is thoroughly domesticated, responds quickly to kind treatment and wins for itself friends and admirers wherever it is known.

CHAPTER II.

Multiplication.

THERE is no animal that has ever been domesticated that is so prolific as the Belgian hare. They will NOT raise a dozen litters a year, as some breeders have claimed through the press, but they will actually produce four nice, hearty litters yearly without harm to the does, or to the vitality of the young. Just take a moment and consider the possibility to such an animal. Poultry does in no way compare with it. If "biddy" lays and hatches one brood of chicks a year she is doing her work nobly. But does she raise the entire brood? It is seldom that she does. Many diseases attack them, and if she brings to a marketable age 50 percent of what she hatched, I believe this is thought to be a good record. I think I can claim, without contradiction, that the mother doe will bring to maturity 95 percent of the young. In fact, if the breeder will do his part only half, they will thrive very nicely, but do not take this for granted that I advise slighting the care of them; just the contrary, the better care and treatment you give them the better results you will obtain. Neither do I want you to construe this to mean that you feed them a half dozen times a day, for

WOULDN'T YOU LIKE TO HAVE THIS BUNNY?

this would be worse than half slighting them. What I want to impress in regard to this is regularity in feeding, but one or two feeds daily is the greatest of plenty, and what you do, do intelligently. Nearly all the troubles arising in the rabbitry are due to ignorance. Ignorance may be bliss, but intelligence is certainly power and also profits, if used in the right way, and this applies to Belgian hares just the same as any other business. This chapter was to tell you the multiplication and addition with the Belgian, and in this connection I will quote from a writer of some years ago:

"Take a doe, for instance, for one year, and she will produce in that time five litters of young; we can assure you of an average of eight to the litter. That would make forty from one doe for the year. Now suppose half of each litter to be does, which is generally the case, your first litter of four does will produce young twice before the end of the year, making sixty-four; added to those produced by the old doe, we have one hundred and four; but that is not all—the second litter of four does, produced by your original doe, will be old enough to produce one litter before the year closes, making thirty-two more to be added to the one hundred and four, making a grand total of one hundred and thirty-six from one doe in a year."

The writer never exactly followed the above method of breeding, but if it was followed there is no doubt but what the results would be almost beyond belief to those not acquainted with the Belgian and their possibilities.

To get the real results a few thoughts might be advised, and it is this: In starting, get GOOD STOCK, and in this I mean all those words imply. You make a mistake by economizing in purchasing your foundation stock, for remember, you can not hope to raise better hares than you buy. Like produces like, and it can not be changed.

After purchasing your foundation stock, the food and care required is precisely the same with GOOD stock as with poor quality, and results far more satisfactory, it matters not if you are raising for fancy or utility. This advice will save you some trouble, but there are a few people who desire to learn everything by experience. Experience is a good school, but often a dear one, and as life is short, I believe it far wiser to profit by the experience of others.

CHAPTER III.

Houses or Hutches for Belgian Hares.

THE writer, having been in the Belgian hare business for over fifteen years, has had considerable experience in building houses and hutches for hares. This article is meant for information to the beginner as well as the veteran hare raiser. No special building is required—a barn, stable or shed, reasonably warm in winter and permitting of good ventilation, but free from direct draughts, is all that is necessary in the way of a building, especially for the beginner. I would not advise elaborate buildings for any beginner; after he has been in the business for a year or two, and is succeeding nicely, then he may plan a building after his own ideas or from those of one who has had years of experience in the business. In fact, I have known beginners to simply obtain a few large drygoods boxes for a few months, and if they are kept dry, and facing the sun in the winter, and under the shade in the summer, this is all that is required till some special building can be erected. Any building can be fitted up very quickly by oneself with the ordinary tools at hand, and using ordinary oak or pine lumber found around most places going to waste.

From my own experience I would always advise hutches to be made eight feet long by four feet wide, and three or four feet high, double deck, or three in the tier, as being the most economical and practical all-round hutches. Such hutches can be used especially for breeding does and their litter until they are six or eight weeks old and are weaned, leaving the doe in the hutch and removing the young to another; or where several litters are put together at once, to a much larger hutch, which will be described later on.

After the writer was in the business a few years, I erected a building especially for Belgian hares, which was forty feet long by eight feet wide and built double deck. The building was set to face the south, and the doors were on the south side, and were made with one-inch poultry netting. The frames were made large, and the netting covered almost the entire front part of each hutch, so as to give the hares the sun in the winter. Then trees were planted on the south of this building so as to give the much-needed shade in the hot days of July and August. The trees are the Carolina poplar, which are quick

One of the buildings of The Pleasant Ridge Rabbitry, Cozaddale, Ohio. This building is 40 feet long, by 8 feet wide, and contains 20 separate hutches. Each hutch is 8 feet long, by 4 feet wide. They are built double-deck plan.

growers, and shed their leaves in the fall so the sun can shine right in the hutches. This building just described has twenty separate hutches, each eight feet long by four feet wide and about four feet high, so anyone can get in nicely. The partitions of the hutches are made partly of wire netting, the lower parts of the lumber, and the top parts with netting, giving each hutch ample ventilation, and still no draughts. Hares do not need artificial heat, therefore no arrangements need be made for it. The fact is, if you attempted to rear hares heated, there is not the least doubt you would lose many by taking colds at the least change of the temperature. Hares should not be kept in a damp place for any length of time at least; dryness and freedom from draughts is the watchword. Each breeding hutch should be provided with a nest box, which can be made from a soap box about eighteen inches long by twelve inches wide and eight inches high. The top should be hinged or made so it will slip on and off at the will of the breeder, with a hole at least six inches on one side for the doe to enter and come out. Hutches should be cleaned out whenever necessary, depending to a great extent on how large the hutch is and how well ventilated. The droppings of hares is valuable for fertilizer, and where many are kept it should be saved, as the fertility it will make will be of considerable value.

Since the writer's business has extended, more room has been wanted to accommodate his stock, and especially so after weaning. Parks have been provided for this, with many separate yards, and while these are nice at certain times of the year, at other times they are not my idea. Hence the con-

STARTING TO THE EXPRESS OFFICE WITH A LOAD OF SHIPMENTS GOING TO ALL PARTS OF THE COUNTRY.

struction of a brand-new building last fall, which I believe the most up-to-the-minute rabbitry in the United States. If anything better has been built it has not come to the notice of the writer, and I aim to scan everything regarding Belgian hares. This new building is forty feet long by twenty feet wide, made with cement floor. It has fourteen separate hutches, eight by ten feet each, also twelve "shipping" hutches. This building has an aisle running through the center four feet wide. All doors are sliding, therefore out of the way, and feeding is a pleasure even when it is raining or during other bad weather. The upper parts of the partition are also made of poultry netting and the lower half of lumber; in this way the hares from one hutch can not see the others, but at the same time there is provided an ideal ventilation without draughts. These hutches are also built double deck plan, but the lower, instead of being three feet high, is between five and six feet, so while one can not stand erect inside the hutch, by bending slightly you have not the least trouble in feeding, cleaning or catching the hares. The upper hutches can be fed nicely from the floor, but of course in cleaning or catching it is necessary to climb a short ladder to get into the hutch. This building is put up in a first-class condition, just as good as the best carpenters knew how to make, and the cost was about $500.00. In erecting these buildings the writer has in mind to stay in the business, but for a beginner no expensive hutches need be made—until he makes a success in the business and expects to continue in it.

The Belgian hare business is on the boom, not the kind of boom there was about fifteen years ago and before the writer entered the business, but a good, healthy demand the year around.

CHAPTER IV.

Feeding the Belgian Hare.

UPON the instructions of feeding, together with housing and breeding, hinges most of the successes and failures in the Belgian hare business. So a close study of these three subjects, with a determination to put into execution the best ideas, should be made. Belgian hares eat anything a sheep will, but they should be fed just as regular as a $200.00 horse or $100.00 cow, with the exception that two meals a day, instead of three, should be given. In fact, the writer has been for the last couple of years feeding his hares but once a day—in the morning—and has secured the best of results. Belgian hares can be overfed just as easily as not fed enough, and the results of overfeeding are far more disastrous. A hungry hare is a healthy hare, has been truly said. This does not mean that they should be only half fed or stunted. Judgment should be used at all times in reference to this business as in any other; if not, better not start at all.

The main foods for hares are oats and clover hay, or alfalfa, which is an ideal ration. A good rule to follow is to give each mature hare a handful of oats once a day, with a little corn added, and some green cut clover or alfalfa hay, at the same time giving fresh water in a small crock holding from a quart to a half gallon, owing to the number of hares in a single hutch. When a doe has youngsters she should be fed more, as you will readily understand that the mother doe does all the work of feeding her young, and in turn she should have extra nourishment to do it with. As soon as the doe has littered she should have from two to three handfuls of oats and most every kind of food in proportion, and as soon as the little fellows begin to show themselves and eat with their mother they should have still more oats and clover hay or alfalfa. Each individual doe must, to a certain extent, be handled differently from her neighbor; that is, she will have her peculiarities and eat more of one thing than another, and her likes as a rule can be relied on. When it comes to feeding green food, it should be fed to the mother doe before the young are born, and by so doing the tendency to hurt the young is eliminated or entirely overcome. This may seem strange to some, but it is nevertheless a fact. You must also have your eyes open when you are feeding your hares each day, for if any hares are moping around in the corner of the hutch and do not come forward promptly when being fed, something is the matter, and an investigation should at once be made, and whatever the trouble may be it should be attended to at once.

The normal condition of the manure is hard and round, and should it at any time have a tendency to be soft and hang together, it shows the hare is out of condition; and while they may come all right by removing all green food for a day or so, if not, a few drops of tincture of iron in their drinking water will be a great aid. In the summer most any kind of green food agrees with them, especially if not given to excess, and none left over to decay or heat. Hares eat weeds just as readily as anything else, but of course never feed any poisonous weeds or vines. In grains they eat oats, corn, wheat, rye, etc. In vegetables, carrots, cabbage, parsnips, turnips, beets, etc. Do not feed potatoes or parings of these, as it does not seem to agree with many of them.

Hares are dainty eaters, and everything around their eating trough, or drinking crock, should be kept nice and clean. Never feed any kind of grain or hay that is musty or moldy in any way. Anyone raising many hares should plan ahead for their food supply both winter and summer. Each evening the writer cuts a large wheelbarrow full of green clover hay just as long as it can be procured, and this is fed to them in the morning with their grain diet. You will notice they will begin on the clover first. In the winter one should have a good supply of carrots, turnips, stock beets or apples to feed them, but from actual observation I have come to the conclusion that carrots are the thing they like best to balance their ration in the winter, when such things as clover or weeds are not available. A small plot of any of these things will be of great value in feeding your hares in the winter. Of course they should be stored where they will keep nice all winter. Cured clover hay can be fed the entire year, and enough can be put in a hutch to last the hares several days, and especially so if a sort of manger is put up so they can not get it under their feet so quickly. Be careful in making your manger so it will not catch their heads and hang them. Some writers have advocated that the hare does not need water, but from my observation all hutch-raised hares need water, and when fed on dry food will drink great quantities of it. True, when they are put out in yards and let to eat green grass as they will, very little water is needed.

Do not neglect to salt your hares at least once a week. This can be done better by putting a large chunk of rock salt in each hutch, and they will eat it as wanted. Any vegetables that have been frozen should never be given to the hares, as it has a tendency to cause them to have diarrhoea and other ailments.

When hares begin to gnaw their hutches, give them a few apple tree limbs and a little salt. Usually this will stop them.

CHAPTER V.

Mating and Breeding.

BELGIAN hares will usually begin to breed at the age of six months, but this is not advisable, for when they are bred so young, their young will not, as a rule, be as large and strong as where you wait till the doe is about eight months old and then breed.

The buck should also be at least eight months old when he is used as a stud buck, and if he is a good, vigorous one he can be used two or three years to does unrelated to him.

In breeding, look to the doe for size and shape and to the buck for color. From a good doe, properly mated, you will be sure to get good youngsters. In breeding, always put the doe in the buck's hutch, and not vice versa. If she is not in heat she will make a plaintive little noise and run from him. After waiting a few moments remove the doe to her own hutch if she is still

A MOTHER DOE AND SIX OF HER BABIES.

unwilling, and try her again the next day and continue until she is served. Nothing is ever gained by leaving the buck and doe together if she will not accept service, as it will only serve to worry the doe in his attempts as well as exhaust the buck. Young does will often not accept service for several weeks. What I mean by young does are those that have never been previously bred. It takes some patience in getting them bred. However, after they have once had a litter of young the trouble on that score is practically at an end. When the doe has weaned a litter of young she will, as a rule, more readily accept service from the buck in from three to five days after. It is not always advisable to breed them so soon after having weaned their young, but this matter should be decided by the condition of the doe at weaning time and the condition of the weather. Does should not be bred so as to litter in July or August, unless the latitude is much cooler than that of Cincinnati. Because a doe refuses service from one buck is no reason why she should not accept from another, but as a rule, when they are really in heat they will accept from any buck. Better results follow from one good service than from several. When the doe is bred, place her in a hutch where she is to raise her family. You may expect the little ones thirty days from date of accepting the buck. We assume the nest box is in its place, and in about ten days before the little ones are due see that the doe is provided with plenty of good fine hay to build her nest with. Many does will make their nest one week to the day previous to having their young. Still others will make their nest earlier, and some later. Should a doe begin to make a nest and she was not bred, it is a sign she is in heat and should be bred at once, if desirable to do so.

After making her nest she will line it nicely with fur pulled from her own body. She may not make immediate use of the extra hay given her, but this need not cause any uneasiness; allow her to pursue her own course and all will go well. Do not fail to provide plenty of good, nourishing food for her during gestation, and see that she is at all times provided with plenty of fresh water, as the thirst of a doe at kindling becomes abnormal, and when not supplied with water have been known to kill their young. After kindling—and for that matter during the entire period of pregnancy—the doe should be kept as quiet as possible. Soon after the little ones arrive it will be well to look in the nest box to see if any are dead; in doing this disturb the mother as little as possible. Some does have been known to leave their young when the nest has been disturbed, but this is an exception rather than the rule.

If any dead ones are found in the nest they should be removed at once. The period of gestation being only thirty days, with an early return of the sexual passion in the hare family, causes many breeders to breed their does too frequently. In order to obtain best results, I would not advise to breed the doe till the young are at least two months old, and in this way you can raise four litters a year and keep your doe in good shape. Too frequent breeding will have a tendency to impoverish the doe, thereby causing the young to lack vigor and strength, which otherwise the doe would have been able to give them. A doe supporting a large litter of young must give forth a large amount of the food she consumes to her young, and it seems almost impossible that a doe can support a litter of a dozen young and make all of them grow as fast as they do. Therefore the breeder should always give the doe plenty of good, succulent food while nursing her young. An ideal ration is green or cured clover hay, oats and a carrot each day, with water. Substitutes can be given, but will hardly do the same good.

There is nothing that I know of which grows so fast as young Belgians which are properly attended to, as they will double their size in a very short time, and this rapid growth continues till they are from six to eight months old, when they grow slower till maturity, which is about from ten months to one year. One buck is able to serve quite a number of does if the services are not too close together. Every breeder should, however, keep two stud bucks so as to furnish stock not related. Do not inbreed, as this will deteriorate your stock; better pay from $5 to $10 for a first-class buck each year than inbreed, for the service he will give you will be worth many times over the price paid for him. It is best to breed the does in the cool of the morning, especially in warm weather. Some authorities claim that an old buck and a young doe beget the largest and best young.

Belgian hares are peculiar in some respects in regards to breeding. Some does will not accept service from one buck, but will from another, and the buck will not be as vigorous as the one she refused service from. Some does will accept two services at once from the same or different bucks, while others positively refuse to accept but one service. It is not all the time that does who accept one or more services are actually with young, still they seem to be in heat according to their actions. Other animals, however, are the same in this respect. Sometimes does will not accept service only under great protest, but they will have a nice, large litter of young.

People who are not acquainted with the habits of the Belgian hares in breeding them often lose patience in getting them to accept service, but after they once have young they generally breed without much trouble in from three days to a week after weaning the young.

Perseverance and patience generally win out in most everything, and breeding and raising Belgian hares is not an exception to this rule.

CHAPTER VI.

How to Care for the Young.

MANY people who are not acquainted with the Belgian hare might think the care of the young of Belgian hares is a difficult job. The truth of the matter, however, is just the opposite.

The mother doe does most all of the work in taking care of her young, so very little responsibility rests on the owner. All that is needed is to give the doe an extra allowance of food, for she will eat considerably more at that time, and her food should be of a milk-producing kind, so she will provide plenty of nourishment for her babies. The writer has found that the best ration that has come to his notice for a nursing doe consists of clover hay, either green or cured, good sound oats, a carrot once a day, with water in plenty. Some little corn can also be given to balance the ration.

People who have been in the habit of raising poultry are much surprised to find that the young of the Belgian hare are so much easier raised and attended to than chickens, and the mortality is next to nothing in the young.

The writer is also a poultry fancier, and as a rule he is satisfied if he can raise to maturity one-half of the chickens that are hatched.

To those who have it handy or can afford the expense, I would recommend that where a doe has a litter of over eight (some have as many as twelve and fourteen), that they give the doe each day a saucerful of bread and sweet milk (not sloppy), about the time the young are making their appearance from the nest box. The milk must be good and sweet, and they certainly thrive on it. Other food such as above recommended should be given at the same time, but of everything you do, do not overfeed the doe or her young, as it is much more disastrous than underfeeding them. There is a saying that a hungry hare is a healthy hare, and this is in most cases true. Cured clover hay or alfalfa should be kept before the doe and young at all times, and you need not be afraid of overfeeding when it comes to hay, but enough can be put in to last two or three days at a time. When it comes to grain and vegetables, etc., do not feed more than they will clean up in a half hour.

The little ones will make their appearance from the nest box in about three weeks from birth. You will find until they are accustomed to their surroundings they will be somewhat timid, but this will soon wear off. The youngsters wil eat the same kind of food their mother eats, but they should

SOME YOUNG BELGIANS GETTING THEIR SUPPER.

also have a little extra, and rolled oats in connection with regular oats is very fine for the young, as they can eat it better than whole oats.

I would not recommend weaning the young from the mother until they are at least two months old, and three months old is better. If you will keep the young with the mother until they are three months old, and in a good thriving condition, you will see they will suffer no backset in weaning, which, when weaned younger, they are liable to have. If they are not weaned until they are three months old they can be put in a large hutch, and the sexes separated at the same time, using aluminum ear tags to mark them with and keep a record of all hares in a book provided for the purpose. In this way you always have a record of every hare, and with but a small amount of trouble a pedigree can be made for any hare sold; that is, providing you are keeping the better class of hares, which in reality is the only kind worth keeping, for if you keep this kind and purchase of a really reliable breeder, they will keep you.

After young hares are weaned they can be kept in a large hutch, or if in summer in a large yard of poultry netting, providing you have a place made so they can get in out of the rain. By giving them large yards or hutches, they will get the much-needed exercise, giving them length of body and a racy appearance. The racy appearance, however, is only necessary when seeking to get good show animals. When hares are raised for the market, their food should consist of more of a fattening nature, and corn should be used, but do not neglect the balanced ration, for Belgian hares like a variety of food just the same as the average human being, and if you give it to them they will soon show the results from it.

From an experience of over fifteen years in the Belgian hare business, I have found that does can be kept together after weaning till they are sold or bred. When bred, each doe should have a separate hutch, about four by eight feet, which is large enough for her and her young till weaning time. I have also found that bucks can be kept together the same length of time without injury to each other. Care must be used, however, not to put a new one in with a lot of others that are over six months old, or some of them will get torn, and probably castrated. When a buck is once removed from a bunch, never put him back among them, as the rest will resent it. If you are sending your bucks to the market and hotels, they can be castrated at from four to five months of age, and will fatten much faster and the meat will be much more sweet and tender. The writer has never practiced that part of the business, as I have never found it difficult to dispose of all I could raise.

After weaning the young from the doe, if the doe is in good condition, she can be bred again in about a week thereafter. Otherwise it is best to wait till she builds herself up a little.

CHAPTER VII.

The Belgian Hare as a Side Line.

THERE are but few people, no matter what regular business or vocation in life they follow, that do not wish for something to revert their mind to at times, and there is nothing better for a "side line" than the Belgian hare.

This is true; it does not matter if you are a lawyer, doctor, preacher, undertaker, teacher, farmer, or wife of one. The Belgian requires so small a space and so little time for a few of them that they are ideal, no matter what your occupation is.

The writer knows the Belgian is adapted to any of the above whose business he has named, and hundreds of others, for he has personally received orders and testimonials from people in nearly every vocation in life saying they were much pleased with the Belgian, and that they could raise them with both pleasure and profit.

After a busy day in the office there is nothing nicer than to come home and take a few minutes out with your "bunnies," feeding them and watching them eat. It is a relaxation of the mind from the more strenuous duties of the day, and attending to them will revigorate the mind as well as the body.

However, a boy going to school during the day may desire to make a little money on the side. Then the Belgian is the ideal animal—they cost but little to start with, and the required space is small and need not be expensive. Belgian hares are bringing good prices at hotels and from commission men, as I have several customers saying they have no trouble in getting twenty cents per pound live weight. If, however, you do not care to dispose of them for eating purposes, then do a little advertising. This is the only way for people to know you have anything for sale. After they know it, then it is up to you to make the sale. It is true that anyone advertising will receive many inquiries from people who are simply inquisitive and would not have the goods advertised even if they were almost given to them free of cost. But if any firm or person were to make a sale for every inquiry he could not begin to supply the demand. Advertising and replying to inquiries gives you a good business education—that is, if you have that business capacity; and it will be a great help to any young man or woman in after life. Each person who replies to your advertisement is different from the previous one, and to a certain extent must be handled differently.

People living in the suburbs of the great cities can not, as a rule, keep chickens, as many neighbors object to them on account of their noise in the morning. There are also objections to raising chickens in close confinement, as they do not do very well and can not do their best when so kept. It is exactly the opposite with the Belgian. No one can make an objection to him, as he is always quiet, unless he happens to be scared, when he will stamp his feet a few times, which will not awaken anyone.

The hare furnishes a nice delicacy for the table the year round, the meat being juicy and tender, and much preferred by some to spring chicken. They should be fried like a spring chicken, when three to four months old, or stewed at six months, or stuffed and roasted like a turkey at eight months old or over.

It is not necessary to feed the hares three times a day. The writer has been feeding his but once a day for several years, and believes they are just as hearty, if not more so, than when fed twice each day.

No one in the country has any excuse for not keeping a few hares, for they will utilize much stuff that would otherwise go to waste, and will turn it to a profit.

Of course, the Belgian does not lay eggs (if he did everyone all over the country would want several), but he has several other good points in his favor against the hen, the main one being that the mother can and will breed and have four litters of young in a year, averaging from six to fourteen to a litter.

Here is what a young man in Massachusetts said not long ago to me in a letter: "I received the hares in good shape and am well pleased with them. Although I attend school, and my time is limited, I enjoy the business very much. I now have about twenty Belgians in all. I also consider the possibilities of the Belgian superior to the hen." What this young man said has been confirmed by hundreds of men, women and children from all over the country.

Anyone in city or country who desires a side line for both pleasure and profit can not do better than give the noble Belgian a trial. He will not disappoint you if you will but do your duty by him.

THREE OF THE BOYS INTERESTED IN RAISING BELGIAN HARES.

CHAPTER VIII.

Diseases and Remedies.

AS a rule the Belgian hare is quite healthy and seldom sick, but even in the best regulated rabbitries there will be, once in a great while, a sick hare to treat. Prevention, however, is better than trying to cure after being sick, and good judgment is required in care and feeding in order to keep them well.

The best advice to offer in this respect is to feed regular, but do not overfeed. Feed nothing but what is in good condition—never any kind of food that is musty or moldy. The hutches should be cleaned whenever needed; and large hutches are always best—four feet by eight feet being about the size for a doe and family till weaned.

Hares need shade in hot weather, sun in very cold weather, and water before them at all times.

The droppings should always be hard and round, and when in any other condition the hare should be looked after at once to ascertain the cause of looseness and should be given treatment.

Only the diseases that are most often found among hares will be given, as a full list would take much time and space, and avail the reader but little.

SNUFFLES.

This is the most common disease, and while but few hares die from it, it is annoying not only to the hare, but to the breeder also. In the treatment of this disease aim to build up the system. Give food that is extra-nourishing; put a little tincture of iron in the drinking water, and the animal will probably be able to throw off the disease. If you feed a mash, put a spoonful of flaxseed in it; and if it is simply a case of sneezing and discharge from the nose, resulting from a slight cold, nothing more in the way of treatment will be required. It might be well to spray the nose gently with luke-warm water, to which a little salt has been added; after spraying wipe dry.

SLOBBERS.

This is another very common disease, the result of indigestion. Use for this a little common salt; simply rub it in the mouth and put a little in the water. A little powdered alum put into the rabbit's mouth will aid in effecting a cure. Change the food, and do not give any green food and but little water for a short time. Almost any case will yield quickly to this treatment. This trouble affects the young after weaning when fed too much green food, but an experienced breeder seldom is troubled. The writer recalls but one experience of this kind; it was his first litter of young, over fifteen years ago. He has learned better how to feed since.

EAR CANKER.

It is seldom this disease occurs in hares, but when it does it usually affects the older hares. Dry scabs form in the ear and it becomes red and much irritated inside. The hare will try to scratch it a great deal. The best remedy is to take one part sulphur and four parts of vaseline or lard, and mix thoroughly. Apply to the inside of the ear with a feather very gently, and go down nearly to the base of the ear, but be careful not to go in too far, as it might cause deafness. It is seldom that more than one treatment is necessary, but if it is, apply as before in a week after the first. A little sulphur is very beneficial to the hares, but only a very slight quantity should be given, as it will cause looseness of the bowels.

ABSCESS.

If an abscess forms let it get ripe, then open with a sharp penknife, squeeze out the matter and wash out the wound with diluted peroxide of hydrogen. Repeat the wash for a few days and a cure should be effected.

WORMS.

Worms in hares are caused by the food and also by the drinking of soiled or contaminated water. Their drinking vessel should be placed where as little of the droppings will fall in it as possible and should be rinsed out thoroughly each day when feeding and watering.

Symptoms: The hares look thin and poorly nourished, although they eat plentifully and ravenously. They are apt to be restless and twitch frequently.

Treatment: Give Santonine, one-fourth grain in grated carrots or in meal mash, three times a day for a few days. Another remedy is powdered sulphur in soft food for a few days.

DIARRHOEA.

This disease is the most deadly that attacks hares and should be looked after at once in order to save the hares.

Symptoms: When feeding, the hare will not, as a rule, come to her food, but will sit drawn up in one corner of the hutch. Examination will show the droppings are loose and sticky, often coming from the hare in a chain. It is mostly caused by feeding spoiled or musty food, too much green food, or that which is partly decayed.

Treatment: Stop all green food, and substitute some nice cured clover or alfalfa hay and a few oats. Drop into their drinking water a half dozen drops tincture of iron to a pint of water. Give them this for at least one week. To entice the appetite, give a sprig of ivy or some ash leaves, both of which have a binding effect on the bowels. If the above treatment does not stop the trouble in a few days, mix some arrowroot in cold water, as thick as it can be given with a spoon; give a spoonful each day till the hare is better, when stop altogether. Breeders claim this will cure when everything else has failed.

PARALYSIS OF THE HINDQUARTERS.

The hindquarters seem to collapse, and the rabbit drags its hind legs uselessly after it, and the muscles of the thighs sometimes wither away. There is practically no cure for this affliction, and the best remedy is to put the sufferer out of its misery at once.

To prevent these diseases feed your hares with as much judgment as your best horse or cow, but understand that once a day is usually sufficient, with exception of breeding does, which may be fed morning and evening. Fresh air, sunshine and shade in the hot summer, plenty of fresh water, and large, roomy hutches is the secret.

CHAPTER IX.

Belgian Hares and the Young Folks.

NEARLY all children and young people, whether they are only six years old or whether they are sixteen, and this includes both boys and girls, desire something to pet—it may be a hen, a cat, a dog, or calf or pony. All these named make fair pets, but how about the Belgian hare? While the Belgian hare will make a "pet," there is something more about him, if raised right, and anyone with ordinary intelligence can do that. There is not only a pleasure in rearing Belgians, but a good profit besides. It does not matter whether one desires to devote himself to the meat propostion or to raise the finest specimens for stock and exhibition purposes.

Lots of people all over the country say they prefer their meat to that of poultry. Now, of course, the Belgian hare does not lay eggs, but what is the old mother doe doing while the hen is laying her eggs? She is rearing her

TWO OF THE HELPERS AT THE PLEASANT RIDGE RABBITRY.

first litter of young, which she brings forth in thirty days after being bred. If the doe is a good one and she is only half cared for, she will raise from six to a dozen youngsters in the first litter. When the youngsters are about two months old the mother can be bred again if in good flesh and vigorous, and there is no use having her any other way, if you will give her any kind of care at all. After she is bred the second time, she will produce just as many young as the first time, and in a good many instances more and better ones. But how about the hen? What is she doing all this time that "bunny" has reared one litter and possibly two litters? Why, the hen has been laying eggs, of course, and she keeps right on till she gets her laying out, when, if she is of the heavier type of poultry she will want to set and hatch her little ones. A large hen will usually set on fifteen eggs, and oftentimes hatch out a dozen chickens, if the rats or other hens don't bother her. Her work is done in three weeks—just a week less than it takes the mother bunny to bring forth her young. But now comes a strenuous time to raise all of those twelve little chickens. Gapes, lice, rats, hawks and a multitude of other things seem to be quite fond of tender young chickens, and if you are not quite careful you will not raise more than a half or third to maturity. But on the other hand, how about the mother doe? It is very little trouble of any kind to raise the young of the Belgian. Why? Because all that is required is to give the mother a little more to eat, give her a little carrot once a day, as it helps produce milk, some good clover hay, or better yet, alfalfa. A little oats and you have a balanced ration for a day, and some does might be given a little corn with the other foods. Such care and feeding as I am suggesting will produce nice, healthy Belgians. But do not overlook one thing essential—it will not produce the stock. The foundation to your herd must be well selected, and there is just as much difference in good and bad Belgian hares as there is in poultry, cows, hogs or any other animal I might name. Beware of whom you purchase. There are just as many rascals in the hare business as any other, taking in consideration the number of breeders. This is simply a "hint."

But to my subject. It is quite rare for a hen to lay and produce more than one setting of chicks in a year. But what of the Belgian doe? She will produce four litters of young in a year, and average about eight to the litter. The loss of young is practically nothing if you keep the doe in good order so she will have milk and can feed her young. She does all this extra work herself, and only asks for a little more food, and the proper kind to produce food for her young.

This doe will produce at least twenty-five young in the course of the year, and I am in touch with breeders who tell me they have no trouble in selling Belgians for eating for eighteen cents per pound; this is to commission men, who in turn must have a profit. Young hares weighing six pounds will at least bring the breeder $1.00 for eating, and if he is raising them for stock purposes, has first-class stock, and really knows they are, he can, of course, get more money for them; but understand, they must be advertised in order to procure the customer. Breeders entering the business can, of course, take their choice. But possibly you do not care to do either, but simply want a few hares to reduce the high cost of living. In the writer's opinion, there is nothing compared with the Belgian that would help more; they take but small room and really do as well, if not better, in confinement than at large.

Most children delight in feeding and caring for them, and if confined in school during the day, nothing is better for the young people than to get out morning and evening attending to the wants of the hares.

Lots of fruits and vegetables go to waste on most every table that could be utilized to good advantage if Belgian hares were raised, and this waste turned into money by giving it to the Belgians. A few hares could live fine from almost anyone's table by giving them the refuse vegetables, etc., that are not spoiled.

Some boys like to roam the country and be on the streets with other boys whose morals may be none too good. If parents would see to it that their children had something to do, when otherwise they might be idle—and there is nothing better to attract them than the Belgian—much of this running around might be averted.

While they are doing this they get an object in life; and if they turn the Belgians into a profit, as well as enjoy the pleasure of their care, they will soon have a nice little "nest egg" in the bank, which will teach them the habit of thrift, and last, but not least, will also help solve the problem, "how to keep the boy and girl on the farm."

CHAPTER X.

Recipes for Cooking the Belgian Hare.

By U. G. Conover, Belgian Editor, and others.

IT is very seldom we hear much said by our readers about cooking the hare, but I am well satisfied that the majority of breeders who have been in the business long know one or several good recipes that would be pleasing and tempting to the appetite.

Most everyone can raise Belgian hares cheaply by feeding them the refuse from the kitchen and garden. But this I do not mean things that are decayed, but fresh things, used promptly and in season.

But here are the recipes, and I would ask that my readers who know more to kindly send me in several. Of course, I do not want theories, but those that have been tried.

FRIED HARE (The Editor's Favorite).

A hare to be fried should be young—not more than six months old, and four months old is better. To kill it, take a small club and hit it a quick lick over the back part of the head, which, if struck just right, will break its neck, killing it almost instantly. In doing this the hare should be held by the hind legs in the left hand and the club used with the right. As soon as the hare is insensible, use a sharp axe and sever the head so it can bleed well. Cut the hare up as you would chicken, not in too large pieces, but previous to cutting up, if possible, let it freeze over night or until it is nearly solid. Anyone having an ice chest can do this. This takes out the animal taste, and, in my opinion, makes better meat.

Before placing the hare in the skillet or frying pan, have your lard hot, and roll each piece in flour before it is placed in.

A hare about six months old should be cooked slowly for from one and one-half to two hours or until tender and a light brown.

If boiling water is poured over it when about half cooked and let simmer until done it will be far better. The seasoning should be done as soon as placed in the frying pan. After the hare is thoroughly done, take out and keep hot until served. A fine brown gravy can be made just the same as made from a fried chicken.

SPICED HARE.

Cut up the hare after it has been very thoroughly cleaned and laid in salt and water for about an hour; pour some vinegar over it and let it remain in the pickle over night; then put a lump of fresh butter about the size of an egg into a deep stewpan, cut up an onion in it, adding one bay leaf, about one dozen pepper corns and part of a celery root; lay the hare in this stew, adding part of the vinegar that the hare was pickled in and salt slightly before stewing. When tender, thicken with flour that has been browned in a spider with butter.

JUGGED HARE.

Cut the hare in pieces; put in kettle and cover with water; add a medium sized onion, with about eight or ten whole cloves stuck in the onion; cover and simmer slowly four or five hours. When tender, thicken with a good tablespoon of flour; add salt and pepper to taste, and just before dishing up pour in half a pint of good port wine. This is rich and good.

STEWED HARE.

Slice four tomatoes, one onion, and one chili pepper into a stewpan and bring to a boil. Cut the hare to pieces and put in as soon as boiling; add one teaspoon of salt and enough hot water to cover. When nearly done, thicken with flour and butter the size of an egg.

BELGIAN HARE STEW A LA AUSTRALIAN.

After the Belgian hare is dressed, put it in salt water for an hour; then place it on ice for twenty-four hours, or in winter hang it out to freeze (like beef, it is better after a day or two in cold storage than fresh).

In preparing it for stewing, cut the Belgian hare in pieces at the joints.

Chop an onion very fine, also a carrot, and put them in a stewpan with a tablespoon of melted butter, and let this brown slightly; then add the meat, which you will have rubbed well with salt and pepper. Let this brown slightly; then add one tablespoonful of flour and let this brown a little. Chop a square inch of ham very fine, almost mincing it, and add to above; then add a little parsley and a few bay leaves minced very fine. Let all this brown nicely; then pour over this a cupful of sherry wine. Cook this for ten minutes, stirring it constantly so it will not burn, and then add enough boiling water to cover the meat. Stir well, season again to taste and let it boil until the meat is tender and serve very hot.

Belgian hare roasted and basted with sherry wine is also excellent.

In dressing the Belgian always keep the liver, as it is the best part of the stew and considered quite a tidbit.

BELGIAN HARE AND WHITE DUMPLINGS.

Joint the hare and soak two hours in salt and water. Take one large onion sliced. Pepper and salt to taste. Make six (or more) suet dumplings. Put all in saucepan with some good stock, and stew slowly two hours. This is a very economical method and produces a really pleasing course.

BELGIAN HARE PIE.

Joint the hare and soak two hours in salt water. Put into saucepan and stew one hour slowly. Make a short paste; line the pie dish with paste; put in the stewed rabbit, with a few slices of bacon, and a few cubes of beef (if liked). Put on the top paste, not forgetting to make small hole in top for steam to escape. Bake in a moderately hot oven.

CURRIED BELGIAN HARE.

Take one hare, two ounces butter, one pint of stock, three onions, one tablespoonful of curry powder, one tablespoonful of flour, one tablespoonful of mushroom powder, juice of half a lemon, half a pound of rice.

Joint the hare; stew with the butter and sliced onions till they are a nice brown. Boil the stock and pour on top of the stewed hare; mix curry powder and flour smoothly with a little of the stock; add mushroom powder and simmer altogether gently half an hour; squeeze in the lemon juice. Serve in the center of a dish, with the boiled rice all round.

A FAVORITE METHOD OF COOKING.

Take one hare, one-fourth pound butter, pepper and salt to taste, one blade of mace (powdered), three mushrooms, two tablespoonfuls of chopped parsley, two teaspoonfuls of flour, two glasses of sherry, one pint of water.

Joint the hare and put in stewpan with the butter, salt, pepper and mace; cook till nearly done; then put in the other ingredients and boil ten minutes, and serve.

JUGGED A LA WILD HARE.

Take one hare, one and one-half pound gravy beef, one-half pound butter, one onion, one lemon, six cloves, pepper and salt to taste, half pint port wine.

Joint the hare; dredge with flour and fry in boiling butter. Have ready one and one-half pints of gravy made from the above beef and thickened with a little flour. Put this into a large-mouthed jar; add pieces of fried hare, an onion stuck with six cloves, a lemon peeled and cut in half, and a good seasoning of pepper and salt. Cover the jar down with a cloth; place it to the neck in boiling water in a stewpan, and stew till the hare is tender, taking care to keep the water boiling. When nearly done, pour in the wine and add a few forcemeat balls. These must be fried or baked in the oven a few minutes before they are put to the gravy. Serve with red currant jelly.

BELGIAN HARE STEWED IN MILK.

Take two young hares, not more than half-grown, one and one-half pints of milk, one blade of mace, a dessertspoonful of flour, a little salt and cayenne.

Joint the hares and put all in a stewpan; simmer very gently till quite tender. Stir from time to time to prevent burning. One-half hour is the usual time for cooking.

A SMALL BUNCH OF YOUNG BELGIAN HARES.

CHAPTER XI.

Why Belgian Hares are More Profitable than Poultry.

By U. G. Conover.

THE following reasons will suggest why it is more profitable to raise Belgian hares than poultry:

Because a Belgian hare will not eat over one-third as much in the same given time as the hen will.

Because most hens do not hatch more than one setting of eggs in one year, whereas a doe Belgian will readily have four litters of young in one year.

Because many young chicks die of the gapeworm and other diseases, seldom over half of them reaching maturity, whereas the doe Belgian will nurse her own young, and if anything like fair care and food are given her, she will raise at least ninety percent of all the young born. It is no uncommon occurrence for a doe to have a dozen young at a litter, raising every one to weaning time, or about three months of age, and then if the breeder does his duty it is rare that one dies.

Belgian hares thrive in all parts of the country—no climate seems too cold or too hot. The writer has customers in Alaska, and one customer's letters say his hares are doing fine, and to prove his claim he sent me some photographs of his hares after having them for one year. The pictures showed that they were in fine condition, and had increased wonderfully, each doe having two litters of seven and nine to the litter, and of the four litters he lost but two of the youngsters. The writer has also shipped hares as far south as Havana, Cuba, and while having no letters telling how they have thrived, I should judge they were all right, as the customer sent for the second lot, amounting in all to nearly a hundred dollars' worth of stock. As for the different locations throughout the United States, Canada and Mexico, the Belgian hare thrives in every climate and altitude.

The Belgian hare has no vermin of any kind, while you can seldom go into a poultry house without getting covered with lice, which are not only in the house, but on the chickens as well, causing them to leave their nests and much other trouble when setting.

Hares even if half cared for have very few diseases, and if a breeder will but give them ordinary care he will raise ninety percent of all born. On the other hand, most of the poultrymen think if they are able to raise fifty to seventy-five percent of their baby chicks to maturity that they are doing quite well. Many farmers' wives even fall below this average. Poultrymen seldom hatch chicks in midwinter, but the doe Belgian can and does have her young when the mercury is fifteen to twenty degrees below zero (the hutches being out of doors right along), and with the very best results to the young.

The Belgian hare stands confinement the best of any animal the writer knows of, a hutch or box four feet by eight feet and about three feet to four feet high is quite large enough for the doe and her young till weaning time. This being the case, persons living in towns and villages, not to mention the country, can keep a trio of breeders where poultry would not be tolerated by the neighbors on account of the noise of the roosters. The Belgian is a very quiet animal.

The up-to-date poultrymen make a few exceptions to some of these statements, saying they are equipped with incubators, brooders, etc., and are not compelled to hatch the chicks by the hen that lays the eggs, etc. This is all right. "Improvement is the order of the age," but those incubators must be watched carefully for three long weeks. A too hot or a too cold temperature is liable to spoil the whole hatch. I think it is an acknowledged fact that the hen can and does hatch a larger percent of her eggs than most of the incubators, which, after all, is but Nature's way. On the other hand, old Dame Nature does everything for the Belgian except feeding and watering once daily.

Possibly some may say it is not necessary to feed the chicks at all, as they "pick their living." This is oftentimes only too true—they pick their living at the expense of the garden, fruit, etc., often damaging as much as they are really worth, if not properly yarded or watched.

As to the profits in hares some people seem to doubt, but the writer has been in the business about fifteen years, and if I were to tell you of the number raised and sold not many would believe it, but keeping a record of each shipment, the kind of stock, price, etc., enables me at the end of the year to know POSITIVELY what has been done.

Supposing you wanted to raise a few for your own eating—the meat is unequaled except, perhaps, by frog legs—many things now going to waste can be turned to good advantage by feeding to the hares (nothing spoiled or decayed, however), and the high cost of living can be greatly reduced, as the cost of meats is the highest of anything we purchase for the table.

If the breeder wishes to raise hares on the commercial basis, then commission men and hotels pay you from fifteen to twenty cents per pound live weight. The number of times the doe breeds in a year, and the number she has at a litter make it a very profitable deal on a commercial basis.

In starting, be careful, as there are swindlers in the hare business as well as other kinds of business. You can not start with a few cents and get reliable stock. Like produces like. Pay a reasonable price for stock, and buy of a responsible man who has a reputation for honesty. If you can not do this, do not start; for if you do you will be sorry and never succeed in the business.

CHAPTER XII.

A Talk With the Beginner in Belgian Hares.

IF you are a beginner in raising Belgian hares, don't be too easily discouraged, as it is not every man, woman, boy or girl that starts in raising Belgian hares that makes a success, or is fitted for it.

After purchasing good stock for your foundation (and any other is worthless, causing disgust and disappointment and dropping out of the business), then it is up to you to do your part to make your Belgians pay. Many beginners do their part well for a time, or until the novelty wears off; then they begin to lose interest and only half attend to their hares, feeding irregularly, either not enough or too much. Any such care as this will not bring you success with hares, or any other live stock, for that matter. True, children more often lose interest after a few weeks or months rather than the man or woman who goes into the business to make a success of it, and to help reduce the high cost of living by furnishing hotels and commission men with hares.

Anyone not determined to do one's part in making the business a success by giving the hares the slight care they must have daily, should not enter the business. While hares do not require but one meal each day, they should get that regular and with judgment, not giving them too much one time and not half enough the next. Such treatment will only tend to make your venture a failure. Many beginners complain that the stock was not good or as represented, when in reality it was not the stock, but the careless way of caring for them. Hares in hutches need some care, as they are unable to "scratch their living" as do poultry when allowed to roam at will.

Supposing you were locked up in a house and unable to get out, you certainly would have to have food and water or else you would perish. Just so with the Belgian. Some people start in the business of raising Belgians expecting to make large profits the first year, which can not be done. No matter if the Belgian does multiply fast, it takes time to get established in any business. It must be worked up gradually. Do not despair—keep at it, as it is the stayer that wins. As the writer pens these words he knows from experience how true those words are. Fifteen years ago he made his start in a small way, each year enlarging and building up his stock. The first and second years he sold but a few dozen, but today in a year he sells thousands, shipping as far as 6,000 miles.

No great business was built up in a year or even a few years. In buying stock the most profitable to purchase is good stock, and breed your stock up, not down, as a good breed of Belgians is always in demand for breeding purposes. Always aim to improve your stock whenever you see a chance to better it, and you will finally reach the top of the ladder.

It does not matter if you are raising stock on the commercial basis, good first-class stock will always pay better than "any old thing" will. Good stock is a little higher in purchasing, but after the first cost, then the cost of feeding and care is just the same with good stock as with that of inferior quality, and the results are far more satisfactory.

Many poultrymen are waking up to the fact that there is money in hares, and are raising them as a side line with their poultry business. The hens lay the eggs and the Belgians produce the meat, which make an ideal combination.

It does not matter what business you are interested in, the Belgian is the ideal animal for the side line, as they are raised for both pleasure and profit. While you are getting one you can not help getting the other.

The writer has customers in about every vocation or business there is— the lawyer, doctor, minister, merchant, postmaster, clerk, poultryman, farmer and laborer, not to mention the wives, sons and daughters of all the above find raising the Belgian hare a profitable recreation. It is an interesting proposition to anyone, and all can make a success of it, if they but do their small part.

Remember that patience and perseverance are required to accomplish all things worth striving for, and it is the stayer who wins, and not the kicker or quitter.

Remember also that man's best capital is his ability to work. The Belgian business properly handled requires some work if you want to succeed and get on top. There is plenty of room at the top, but do not wait for the elevator—climb the ladder before.

CHAPTER XIII.

The Belgian Hare Business and Its Outlook.

HE writer is in a position to know something of the outlook of the Belgian hare business, being Belgian editor of The American Poultry Advocate and in correspondence with hundreds of breeders all over the country. The outlook for the Belgian hare business is better and brighter than it has been for the last fifteen years, or since the writer entered it.

Naturally, anyone contemplating entering any business desires to know the chance of making something on his investment and labor. While the writer knows scores of breeders all over the country who are making consid-

A MOTHER AND HER LITTLE FAMILY.

erable money with hares simply as a side line, speaking for himself, last year was his banner year in the business, and if he was to tell the amount of business he did in 1914 there are not a great many people who would believe it, and especially those not acquainted with the business.

The writer has contended for several years that anyone could take a doe Belgian and with proper care and feeding, etc., could make much more profit on the doe than anyone could do on the finest cow in the country. He has tested this thing out to his entire satisfaction, and as his home town is a dairy country, with a large creamery where the dairymen take and sell their milk, he has talked to many people in the dairy business and knows just what can be made by one's best efforts with good cows fed right.

Now, I would not want any prospective hare raiser to get it in his head that he could purchase one or several Belgian hares and feed them "occasionally," and could get independently rich; neither can you with cows or sheep or poultry.

While the culture of Belgian hares is very light, even old men eighty-nine years old can, and do, make a fine thing out of them, it must be understood that while only one meal a day is necessary, it should be given with regularity and with judgment. Hares are in reality small eaters—even the most hearty of them are—and without the least exaggeration, one hen will eat three times as much grain as a Belgian will, that is, to give each as much as they want or are satisfied with.

The only thing that is necessary to make a success of this business is to have ordinary intelligence and enough "push" to see that the hares are attended to properly. It is quite true also that a beginner will make some mistakes, but if he will go at it right he will make but very few, and by RIGHT I mean in the beginning to purchase some good book on the subject, and get some idea about it, and right here I would like to say in this connection that the writer has just had a brand new book published entitled "A B C of Belgian Hare Culture." It is really an encyclopedia on the Belgian hare, and has twenty-two chapters, fifteen halftones from life, with nearly 30,000 words. The price of this book is but fifty cents, and to anyone entering the business it is worth its weight in gold.

I would not advise anyone contemplating entering this business to purchase several hundred dollars' worth of stock to begin with. A trio—two does and a buck—is an ideal start; and I think nothing so good as a breeding trio, that is, two does bred by the breeder to his best bucks, and a buck sent along not related to the does or their prospective young. If no one cares to invest the price, which is very reasonable considering the possibilities, they should not enter the business, even as a side line.

I also wish to caution beginners in "starting right." Now there is a right and wrong way to start in this business, as well as any other, and if you start wrong the chances are you will give up in disgust, with no success, when the real trouble is with yourself.

There is a class of people who think you can start with the best stock ever raised for about $1.00, with the possibility that the breeder would pay the express, too. There is just as much difference in the quality of Belgian hares as there is in hogs, cows, horses, poultry, etc. Just a couple of days ago a young man came to me saying he had had no success with his hares. After a little questioning I found he was one of the deluded kind who thought he could start for a "dollar" and make the business pay. He had made a complete failure because he had not started right. One does not need exhibition stock, costing $50.00 per trio to begin with—far from it; a trio costing $15.00, however, is necessary, and if you pay much less you will get something not even worth the express charges, let alone the feeding. Everybody is fond of a "bargain," but you should remember that a bargain is not what you pay, but what you get. Price is but one of the things to be considered. If you fail to get the quality you throw away your money, no matter how small an amount you pay. If everyone would only keep this in mind at all times in purchasing, there would be less poor, worthless goods sold and the public would have more money.

In starting with GOOD stock, the first or initial cost is but some little higher than when purchasing "any old thing," but after the purchase of live stock then the care and feeding is precisely the same; but the one starting right has the advantage, and his results will be far more satisfactory. It is an utter impossibility to raise the best stock in the country if you have purchased the poorest. Like begets like, and Nature does her work well. Therefore, if you are thinking of trying your hand at the business as a side line to help you out in the present high cost of living, if you can not afford to pay a reasonable price for your foundation stock, then don't start at all.

I believe from reports from breeders all over the country that those starting right are doing a nice business, and as to my own, will say my mail brings me daily from forty to fifty letters making inquiry for the FINEST HARES THAT GROW.

CHAPTER XIV.

The Record and How to Keep It.

MANY breeders are at a loss to know how to go about it to keep a record of all their stock. This is not at all difficult if gone about in a business way. When you purchase good hares (and all others are worthless), get the pedigrees for them, so when you sell stock you can also give their pedigrees to your customers. The next step is to obtain hutch cards for each breeding doe, and keep a record of her breeding, testing, kindling, weaning the young, etc. These cards are inexpensive, costing but ten cents per dozen.

As soon as you wean the young you need ear tags to mark each hare so you will know exactly his age, breeding, etc. These tags are made of aluminum, are very light and have numbers on them, and are put on with a poultry punch. The tags and punch cost but very little and each breeder should have a supply on hand.

Last, but not least, you should have your ledger with all this information in so you can at once refer to it, and know what disposition you have made with each hare, and whether kept or sold. A blank book for this purpose will cost but little, and "what is worth doing is worth doing well," or at least as well as you know how.

While a "pedigree" in itself does not make a fine hare, it is a fact well known among the best breeders that hares with pedigrees are, as a rule, far better stock than those who have none, for the simple reason that a poor stock is not worth recording a pedigree. The only remedy is to purchase hares from a reliable breeder, and whether a certain breeder is reliable or not is ascertained by their old customers. It is not often a person will praise stock or breeder if he is not entirely pleased with the purchase.

CHAPTER XV.

The Pleasant Ridge Rabbitry.

THE author of this book desires to tell the readers about The Pleasant Ridge Rabbitry, at Cozaddale, Ohio.

Cozaddale is located in Warren County, Ohio, thirty-one miles from Cincinnati, on the B. & O. S-W. Railway, and the rabbitry is one mile east of the station, this being the postoffice, express and freight office. All shipments made from this rabbitry are from this place.

The Wells Fargo & Co. operate the express business at this point, and while they have offices in all parts of the United States and in some foreign countries, to points where it does not reach, it transfers its business to other express companies who do deliver at points where customers are located.

A new building of The Pleasant Ridge Rabbitry erected in 1911. It is 40 feet long by 20 feet wide with cement floor. Has 14 separate hutches 8 by 10 feet each, also 12 "shipping" hutches.

It has an aisle running through center of building 4 foot wide, with all sliding doors, and the upper half of partitions of inch poultry netting. Hutches are built double-deck and it is a first-class building costing about $500.00.

Under the ruling and rates put in force by the Interstate Commerce Commission on February 1, 1914, it does not matter whether your shipment goes over one or several express companies; the rate is just the same to that certain point. This is quite a help over the old method.

The rural free delivery passes my door each weekday and delivers and collects all mail, so that most all mail is dispatched quickly, especially where stamps are included for postage.

The writer received as many as sixty letters and postals in a single day, with an average of forty each day in the spring season, and this requires some time and work to reply to them. Being the Belgian Hare editor of the American Poultry Advocate, Syracuse, N. Y., makes my mail still heavier.

I have but one day a week to make shipments of orders for Belgian hares, this being on Monday. This gives the hares a whole week to reach destination, and if not delayed can make California by the following Saturday. To points like Alaska it requires eighteen days. The writer, having made shipments to Alaska, can specify to a certainty the number of days. On the other hand, shipments to Cuba require only about a week. In the spring season I make as many as sixteen shipments in a single day, and I keep this up for many weeks, then down to ten and twelve shipments in a day. There is seldom a Monday passes but what a half dozen shipments, at least, go out to my customers located everywhere.

It is well known by anyone at all posted in the Belgian hare business that the FINEST HARES THAT GROW (trademark copyrighted) are raised and sold at this place. This is not simply a name adapted by the proprietor, but it is indorsed by thousands of pleased customers located everywhere. The documents are on file for proof, and a few of the thousands are shown in the "EVIDENCE," "THE VERDICT" and "POSITIVE PROOF." They are excellent reading for anyone interested in Belgian hares, and especially the better grade ones, as the proprietor will not raise and handle "the any old thing."

Do not be stampeded by advertisers offering hares below cost, or lower than for market purposes. When commission men pay as much as $1.00 for hares four to five months old for eating, do not hesitate to pay a few cents more for the best hares for breeding purposes. Remember that QUALITY AND PRICE GO HAND IN HAND. Can you purchase as good a horse for $50.00 as you can for $150.00? Not by any means. So it is with Belgian hares; you can not purchase as good a hare for $1.00 as you can for $5.00.

Twenty breeding does are kept the year round in a special building, forty feet long by eight feet wide, put up for this particular purpose. I can not nearly supply the demand with these breeders, and much stock is purchased from my customers raising them for that purpose.

Another large building is erected with a cement floor, twenty by forty feet, double deck, with sliding doors, and fourteen large rooms in same, and many small ones. This building is used for hares after weaning and until sold. This is an ideal building for this purpose, and other buildings are also used for the hares.

It was stated by a New York drummer that he had traveled over the country from New York to St. Louis, and that the Pleasant Ridge Rabbitry was the best equipped that he had seen, and he made it his business to see all he could hear of, as he was interested in Belgian hares.

Visitors who really mean business are always welcome at the Pleasant Ridge Rabbitry, but those people who are out sightseeing and simply inquisitive in regard to the business I can not give much of my time, as the many duties I am compelled to look after will not allow. It is not necessary to come in person to purchase Belgian hares, as usually my largest customers are at a considerable distance. Anyone expecting to purchase exhibition stock, if near, should call and pick them out personally; still if they do not come up to specifications they may be returned.

One last word. Do not make the mistake in purchasing cheap, good-for-nothing stock, which would not be worth the express charges. After the first cost, which is a little more, then the cost of feeding and care of poor stock is precisely the same as with GOOD STOCK, and the results are far more satisfactory with THE FINEST HARES THAT GROW.

Many of my customers have won first prizes at such shows as Madison Square Garden with stock purchased of me with much competition against them. I do not hesitate in putting my exhibition stock against any in this county.

CHAPTER XVI.

The Profit in Raising Belgian Hares.

THE question has often been asked me, "Is there really any profit in raising Belgian hares?" The writer has been in the business over fifteen years, and it is needless to say if I did not find a "profit" in raising Belgian hares I never would have stayed with it this long. There is more of a chance for a man or woman with a few dollars to make a nice living and lay some money up besides than in any other business with the same investment. If you doubt it, bring on your proof of any business with, say, an investment of $25.00 or $50.00 that will bring you at least 100 percent profit. That sounds big, but I have been at it long enough to know what I am talking about.

There are several ways one can make Belgian hares profitable—you may either sell as breeding stock, and this requires advertising, correspondence, printed matter and sometimes a stenographer, if your mail is a very large one. The other way is to raise for commercial purposes, selling your hares to hotels, hospitals, commission men, etc. There is not nearly the work in selling this way; still you can not get as much for a hare for eating as you can for breeding; besides for breeding purposes, no breeder should raise and try to sell but the best grade stock, for they will bring better prices, and the cost of care and feeding is just the same as with the inferior stock. I do not care if you expect to raise them for commercial purposes alone, the best stock will always bring you more money than the poor scrub stock will.

Breeding hares already bred sell all the way from $6.00 each as to high as $20.00 for exhibition stock. Hares for hotels will bring about twenty cents per pound or $1.00 for about a five pound hare.

In selling to hotels, etc., all you have to do is to attend to your hares. But in selling breeding stock, your correspondence is the hardest of your work connected with the business.

Some prospective customers say they do not see the Belgian hares quoted in the open markets. This is quite true, for the number raised is not sufficient to make a quotation. When a commission man can only obtain hares once in a great while they will not be quoted. However, write the commission men and they will gladly take your hares whenever you have a supply to offer them. Commission men in New York City have been known to telegraph to their customers who raise hares, asking them to ship them in. Still, you would not see a quotation in the markets for them.

The Belgian hare business is just as profitable as poultry raising, and personally I have always claimed it was much more so, as I have the figures in my possession to prove it, but where one man will succeed with the hare the other man might do better with the hen. So it is up to the man or woman who goes into the business to get the money out of it. It's there, all right; plenty of it, and there will be plenty of it as long as people must eat.

The nice part of the hare business is that they multiply so fast. You can breed a doe four times a year, without harm to herself or her prospective young, raising from twenty-five to forty hares from a single doe in a year; can you raise anything near as much from a hen or several hens?

One does not need to drop everything else when he starts in raising Belgian hares. Simply go at it as a "side line." Nothing I know of is any better for that. Then when you see you are succeeding, make it a main line, instead of a side line.

If you do not care to PUSH the business, you will not succeed to a large extent; this applies to every business you may enter. If you do not take care of your business it will not take care of you. Try the Belgian hare, start right, push it with vigor and see the results.

CHAPTER XVII.

American Standard of Excellence.

DISQUALIFICATIONS: 1. Lopped or fallen ear. 2. White front feet or white bar or bars on same. 3. Decidedly wry front feet. 4. Wry tail. A specimen should have the benefit of any doubt.

COLOR.—Rich rufous red (not dark, smudgy color) carried well down sides and hindquarters and as little white under the jaws as possible—20 points.

TICKING.—Rather wavy appearance and plentiful—15 points.

SHAPE.—Body long, thin, well tucked-up flank, and well ribbed up; back slightly arched; loins well rounded, not choppy; head rather lengthy; muscular chest; tail straight, not screwed, and altogether of a racy appearance—20 points.

EARS.—About five inches, thin, well laced on tips, and as far down outside edges as possible; good color inside and outside, and well set on—10 points.

EYES.—Hazel color, large, round, bright and bold—10 points.

LEGS AND FEET.—Forefeet and legs long, straight, slender, well colored and free from white bars; hind feet as well colored as possible—10 points.

SIZE.—About eight pounds—5 points.

CONDITION.—Not fat, but flesh firm like a racehorse, and good quality of fur—5 points.

WITHOUT DEWLAP—5—Total—100 points.

www.ingramcontent.com/pod-product-compliance
Lightning Source LLC
Chambersburg PA
CBHW062206220526
45470CB00009B/2935